IDEA

CREATIVE

SNEAKY PRESS

A catalogue record for this work is available from the National Library of Australia.

ISBN 9781922641069

Sneaky Press
Melbourne, Australia.

# The Book
# of
# Random Brain Facts

Sneaky Press

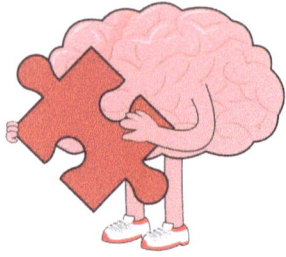

# Contents

Random Facts about the Human Brain    p.6

Random Facts about Animal Brains    p.12

Random Facts about Studying the Brain   p.14

Old Myths about the Brain    p.18

More Random Brain Facts    p.20

Brain Idioms    p.22

Brain Jokes    p.26

Brain Teasers    p.28

Brain Teasers Answers    p.30

# Random Facts about the Brain

The average human brain is 167mm long, 140mm wide and 93mm high.

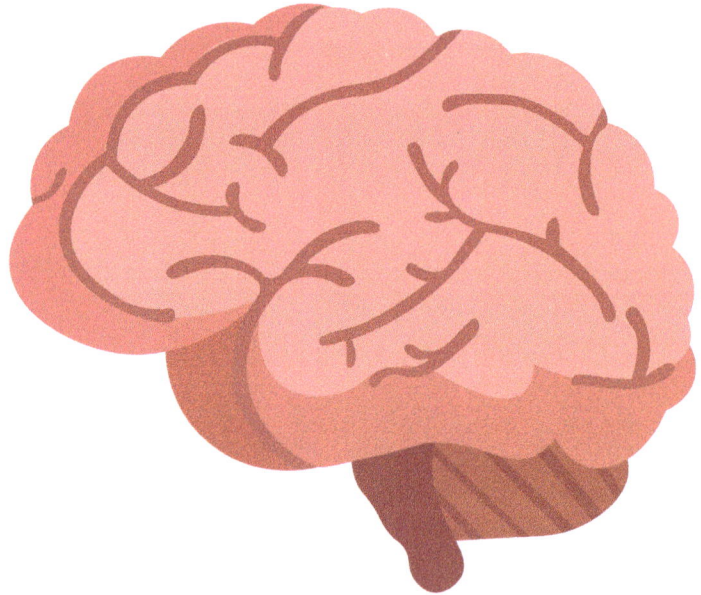

**Did you know?**

Most people have about **70,000** thoughts each day?

That is a whole lot of thinking!!!

The human brain has about 100,000,000,000 (100 billion) neurons – yes, 1 with 12 zeros.

In comparison, a fly's brain contains only 337,856 neurons – approximately 0.0003 % of the number of neurons in a human brain.

The human brain triples in size in the first year of life.

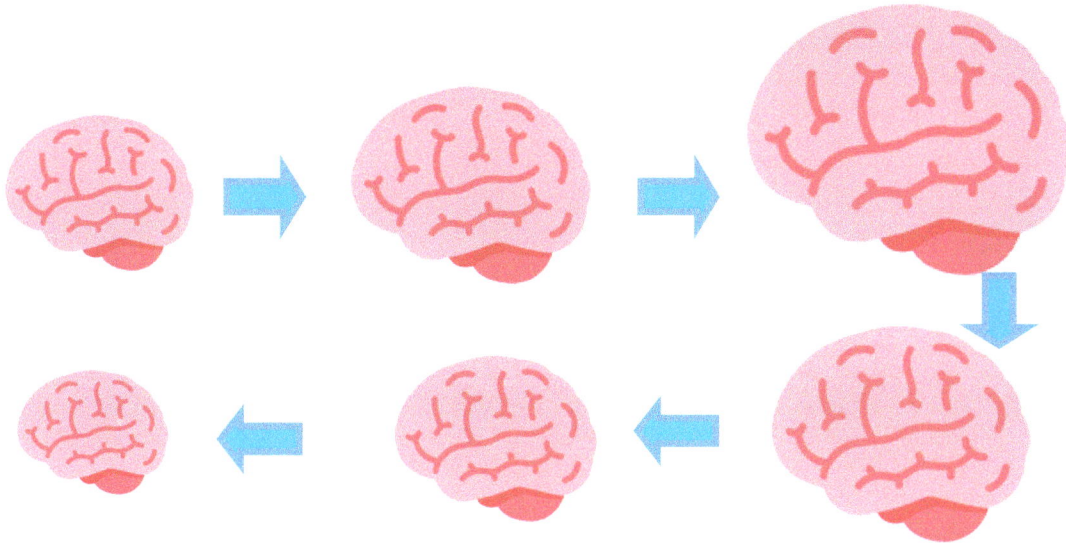

The brain shrinks a quarter of a percent (0.25%) in mass each year after the age of 30.

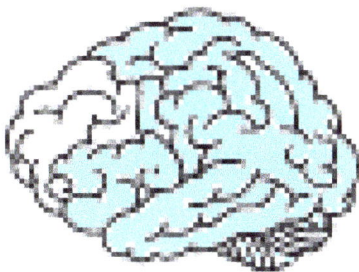

About 75 percent of the human brain is made up of water.

Average

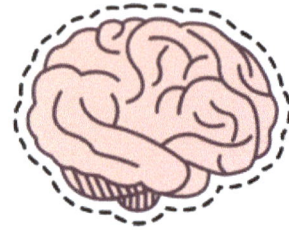

Smaller
than
Average

The heaviest human brain ever recorded weighed about 2300 grams. The average brain weighs about 1400 grams.

The brain of the great physicist Albert Einstein weighed 1,230 grams.

The whereabouts of Albert Einstein's brain was unknown for over 20 years.

The pathologist who did the autopsy stole it and kept it in a jar.

A human brain uses less power than a refrigerator light each day – 12 watts of power.

That is the same amount of energy contained in two large bananas. Even though this may seem very energy efficient, it is an energy hog.

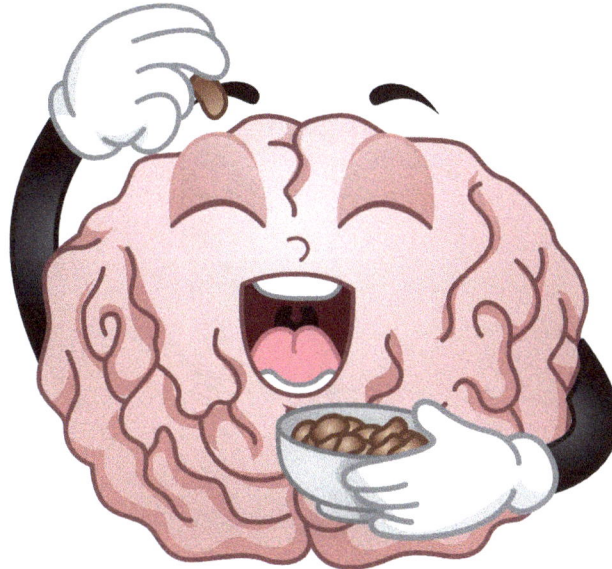

It is only 3 per cent of the body's weight but consumes 17 per cent of the body's total energy. It also uses 15-20% of the body's oxygen supply.

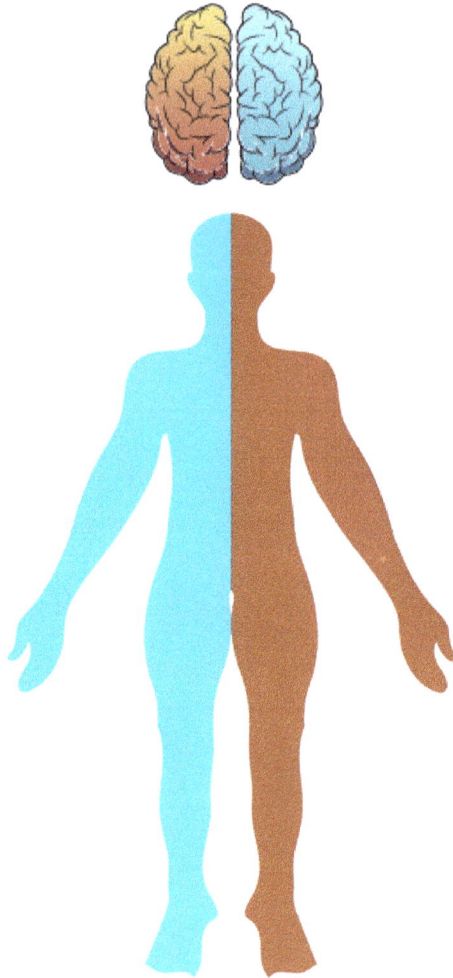

The right side of the brain controls the left side of the body, and the left side of the brain controls the right side of the body.

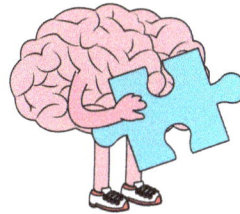

# Random Facts about Animal Brains

The brain of a worker honeybee only weighs approximately 1 milligram.

The brain of an adult koala weighs approximately 19 grams.

An average domestic cat brain weighs approximately 30 grams.

The brain of a great white shark weighs less than 45 grams. Almost 20% of this small brain for such a large creature is devoted to the sense of smell.

The average brain of a killer whale weighs approximately 5,000 grams.

An average elephant brain weighs approximately 6,000 grams.

The animal with the largest brain is the sperm whale. It weighs approximately 9,000 grams.

The oesophagus (the part of the body that connects the mouth to the stomach) goes right through the brain of an octopus.

# Random Facts about Studying the Brain

The study of the structure of the brain (and nervous system) is called neuroscience.

Psychology is the study of how the brain affects behaviour.

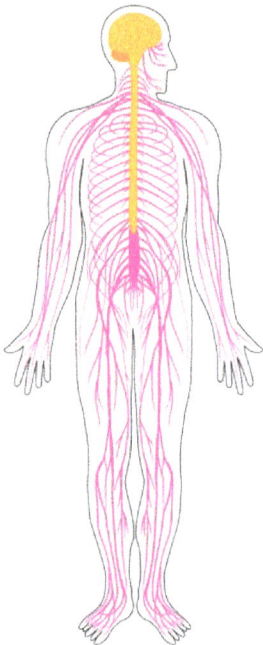

The brain is part of the central nervous system which also consists of the spinal cord. (the yellow in the image on the left.)

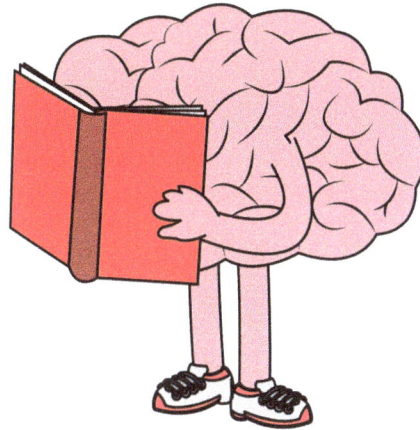

There are over 7,000 brains in a Brain Bank at Harvard University used for research.

There were successful brain surgeries as far back as the Stone Ages. That is at least 4000 years ago.

"Brain freeze," the headache you sometimes get when you eat something cold has the scientific name of "Sphenopalatine ganglioneuralgia".

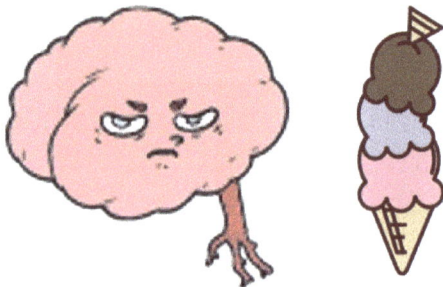

Electrical activity in the brain was first recorded by an Electroencephalograph in 1875.

The brain produces a range of brain waves depending on how alert a person is.

When you are awake and alert, your brainwaves are small and frequent — these waves are called Alpha waves.

When you are almost asleep, your brainwaves are taller and a little slower — these waves are called Theta waves.

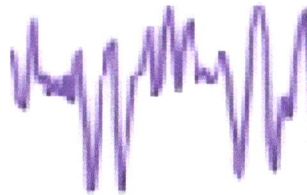

When you are in deep sleep, your brainwaves are at their tallest and slowest — these waves are called Delta waves.

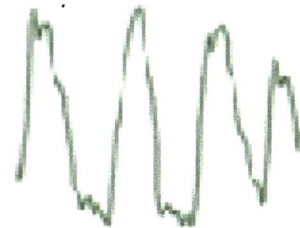

# Old Myths about the Brain

Insomnia (not being able to fall asleep) could be cured by placing a goat's horn under a person's head while they slept.

Anxiety caused by bad dreams would disappear if a person told the sun about their dreams.

Rubbing baby teeth with the brain of a rabbit is an old folk remedy believed to prevent tooth decay.

Ancient Greek philosopher Aristotle believed that the brain was a cooling device for the human body.

# More Random Brain Facts

On March 4, 2001, neurosurgeon Dr. Scott R. Gibbs inflated a 9-story tall hot air balloon shaped like a brain for the first time.

The Dalai Lama keeps a plastic model of the brain on his desk at home.

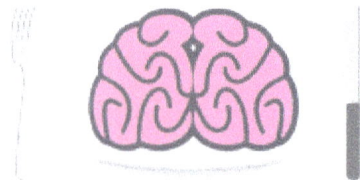

There are over 100 movies with the word brain in the title.

In the 3rd century, the Roman emperor Elagabalus was rumoured to having been served 600 ostrich brains at a single meal.

Minerva was the ancient Roman goddess of wisdom and war. She was the daughter of Jupiter and was born when she leaped from Jupiter's brain, an adult dressed in armour.

Egyptians would usually remove the brains through the nose during the mummification process.

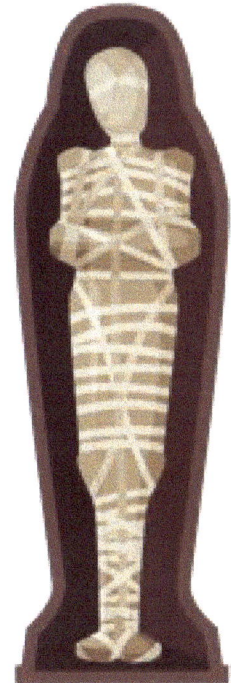

The first recorded use of the word brain was written about 1,700 B.C.E.

William Shakespeare uses the word brain 66 times in his plays.

# Brain Idioms

To have brain like a sieve means to have trouble remembering things.

To be all brawn and no brains means to be very strong but not very clever.

To have something on the brain means you are constantly thinking or talking about something.

A no-brainer is a decision that you can make very easily because the best option is so obvious.

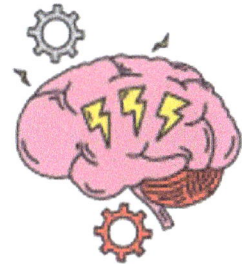

To be the brains behind something is to be the person who planned or organised everything.

To pick someone's brains means to ask someone questions about a specific subject to obtain advice or information.

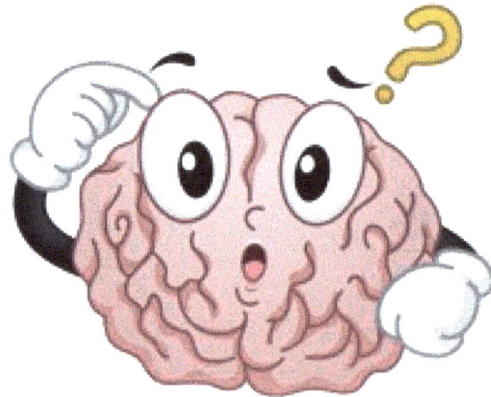

To rack one's brains means to try very hard to think of or remember something.

To wrap your brain around something means to concentrate on something to help you understand it.

# Brain Jokes

What does a brain do when it sees a friend across the road?

Gives a brain wave.

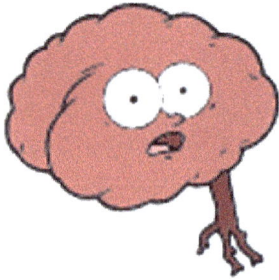

How are brains similar to sponges?

They both like to soak stuff up.

What do you call an empty skull?

A no brainer.

What did the doctor say to the person with a hippopotamus sitting on his head?

It looks like you have a lot on your mind.

Why didn't the brain
want to shower?

It didn't want to be
brainwashed.

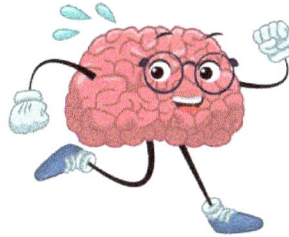

Why did the
brain go for a
run?

It wanted to
jog its memory.

In what kind
of storm does
it rain brains?

A brainstorm.

How do you change a brain into water?

Remove the 'b'.

Why do brains sleep
with lollies under their
pillows?

So they have sweet
dreams.

# Brain Teasers

1. I get wetter the more I dry.

   What am I?

2. I have a face and two hands, but no arms?

   What am I?

3. I go up every day. I never come down.

   What am I?

4. I have many keys, but I can't open any lock.

   What am I?

5. I have a thumb and four fingers, but I am not alive.

   What am I?

6. I am full of holes, but I can still hold water.

What am I?

7. I follow you and copy your every move, but you can never touch me or catch me.

What am I?

8. I am a building with thousands of stories.

What am I?

9. The more you take away from me, the bigger I become.
What am I?

10. The more you take of us, the more you leave behind.

What are we?

# Brain Teaser Answers

1. A towel
2. A clock
3. Your age
4. A Piano
5. A glove
6. A sponge
7. Your shadow
8. A library
9. A hole
10. Footprints

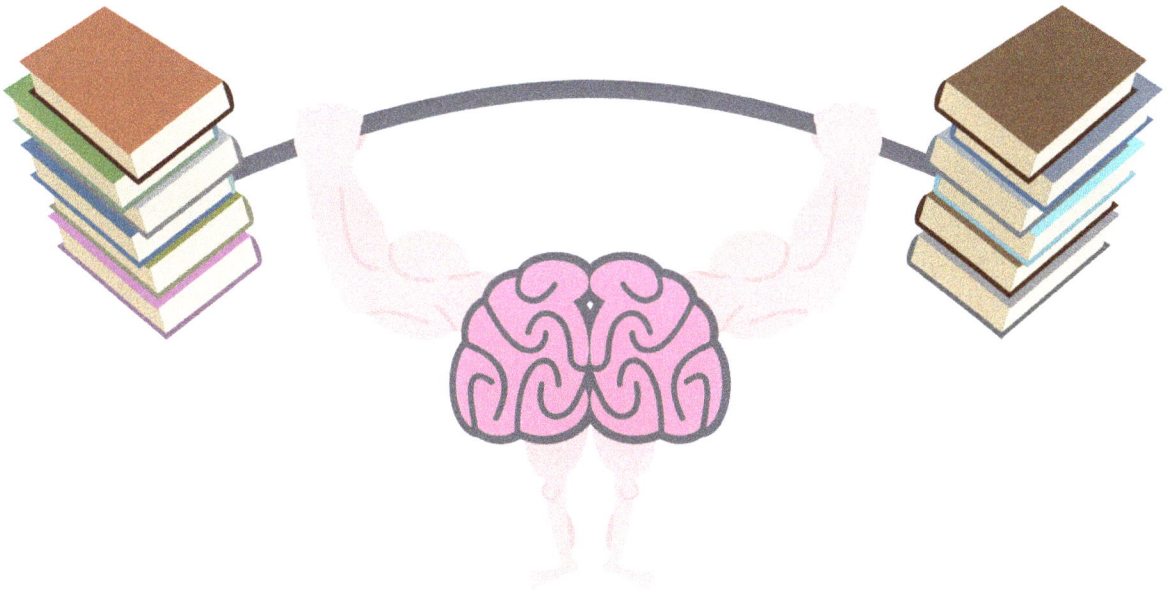

www.ingramcontent.com/pod-product-compliance
Lightning Source LLC
Chambersburg PA
CBHW041911220326
R18017400001B/R180174PG41597CBX00004B/1